T0378280

SPOTLIGHT ON OUR FUTURE

PROTECTING OUR EARTH'S LAND

GENE BROOKS

PowerKiDS press™

NEW YORK

Published in 2022 by The Rosen Publishing Group, Inc.
29 East 21st Street, New York, NY 10010

First Edition

Editor: Theresa Emminizer
Book Design: Michael Flynn

Photo Credits: Cover Hero Images/Getty Images; (series background) jessicahyde/Shutterstock.com; p. 5 (all) Jeremie Richard/AFP/Getty Images; p. 6 Thammanoon Khamchalee/Shutterstock.com; p. 7 Joao Laet/AFP/Getty Images; pp. 8–9 Everett Historical/Shutterstock.com; p. 11 Mohamed Abdulraheem/Shutterstock.com; p. 12 Dogora Sun/Shutterstock.com; p. 13 FloridaStock/Shutterstock.com; p. 15 zeljkosantrac/iStock/Getty Images; p. 16 Sean Pavone/iStock/Getty Images; p. 17 Shane Gross/Shutterstock.com; p. 18 Joecho-16/iStock/Getty Images; p. 19 a4ndreas/Shutterstock.com; p. 21 Stephane Cardinale/Corbis/Getty Images; p. 22 chudakov2/iStock/Getty Images; p. 23 Elena Elisseeva/Shutterstock.com; p. 24 Idoho Statesman/Tribune News Service/Getty Images; p. 25 Graeme Sloan/AP Images; p. 27 Philip Rozenski/Shutterstock.com; p. 29 oneinchpunch/Shutterstock.com.

Library of Congress Cataloging-in-Publication Data

Names: Brooks, Gene.
Title: Protecting our Earth's land / Gene Brooks.
Description: New York : PowerKids Press, 2022. | Series: Spotlight on our future | Includes glossary and index.
Identifiers: ISBN 9781725324305 (pbk.) | ISBN 9781725324336 (library bound) | ISBN 9781725324312 (6pack)
Subjects: LCSH: Environmental quality--Juvenile literature. | Environmental degradation--Juvenile literature. | Global climate change--Juvenile literature. | Clean energy--Juvenile literature. | Marine ecosystem health--Juvenile literature. | Deforestation--Juvenile literature.
Classification: LCC GE140.5 B76 2022 | DDC 363.7--dc23

Manufactured in the United States of America

Some of the images in this book illustrate individuals who are models. The depictions do not imply actual situations or events.

CPSIA Compliance Information: Batch #CSPK22. For further information contact Rosen Publishing, New York, New York at 1-800-237-9932.

Find us on

CONTENTS

CHAPTER ONE

EARTH'S CHANGES

In 1910, Glacier National Park in Montana had more than 100 glaciers. Today, there are fewer than 30 glaciers left there. Scientists think that within another 30 years, almost all of the park's glaciers will be gone.

Why are the glaciers disappearing? It's because of a general warming trend on Earth that's part of a problem called climate change.

About two-thirds of Earth's fresh water is in glaciers. When the ice melts, this water is released into rivers, streams, and oceans. This could make sea levels rise as much as 200 feet (61 m). Coastal areas and major cities could flood.

Climate change is a serious problem, but there's reason for hope. Everyday people can still help the planet. Earth is our home and we must all protect it.

Okjokull (Ok) glacier in Iceland once covered 6 square miles (9.6 square km), but it's gone now. This marker shows where the glacier once stood.

Bréf til framtíðarinnar

Ok er fyrsti nafnkunni jökullinn til að missa titil sinn.
Á næstu 200 árum er talið að allir jöklar landsins fari sömu leið.
Þetta minnismerki er til vitnis um að við vitum
hvað er að gerast og hvað þarf að gera.
Aðeins þú veist hvort við gerðum eitthvað.

A letter to the future

Ok is the first Icelandic glacier to lose its status as a glacier.
In the next 200 years all our glaciers are expected to follow the same path.
This monument is to acknowledge that we know
what is happening and what needs to be done.
Only you know if we did it.

Ágúst 2019
415ppm CO_2

CHAPTER TWO

FORESTS IN DANGER

Forests cover about one-third of the planet. They provide food and shelter for many animals. Forests are important for humans too. We use trees to get wood to build our homes. We eat the fruits and nuts that they grow. We even use some plants and trees to make life-saving medicines.

Trees produce oxygen while absorbing, or taking in, carbon dioxide. This is important for the health of the planet. Carbon dioxide is a **greenhouse gas** that contributes to the **greenhouse effect**.

In 2019, large parts of the Amazon rain forest burned. The fires may have been set to clear land for human use.

Unfortunately, people have destroyed large areas of forest around the world. The act of cutting down forests is called deforestation.

Deforestation adds to the problems of global warming. Fewer trees means more carbon dioxide in the atmosphere. Carbon dioxide holds in heat from the sun, causing temperatures to rise around the world.

CHAPTER THREE

THE STORY OF SOIL

Soil and rocks go through natural processes of breaking down and being moved to new places. Wind, water, and ice cause these processes naturally. However, deforestation and overfarming break down soil unnaturally fast.

During the 1930s, land overuse caused a huge dust bowl in the United States. Thousands of families had to leave their farms on the Great Plains.

Agriculture also takes **nutrients** from soil. In nature, nutrients are returned to the soil when dead animals and plants break down. However, if farmers use the same field many times, the nutrients in the soil will run out. The field will become useless. Empty fields are exposed to wind and rain. The water and wind wash and blow away loose dirt and soil. This can create areas known as dust bowls. If this happens, many plants and animals can't find the food or water they need to live. Today, this sort of damage has reduced the productivity of land around the world by 23 percent.

9

CHAPTER FOUR

DISAPPEARING OCEAN SPECIES

Humans have made large changes to about three-quarters of Earth's land **environments**. The same can be said about 66 percent of **marine** environments.

As a result of these changes, more species are becoming extinct, or dying out completely. Today, overfishing puts ocean species in danger. About 3 billion people around the world eat seafood regularly, but poor fishing practices are driving some species into extinction. Overfishing harms marine **ecosystems**, and that harms other ecosystems.

Oceans cover about 70 percent of Earth's surface. Marine plants produce more than half of the world's oxygen. At least 15 percent of Earth's species live in the ocean. Damage to these species can harm not only the animals but also the entire ecosystem around them.

Sea turtles can become caught in fishing nets. They often die because of this.

CHAPTER FIVE

DISAPPEARING LAND SPECIES

Species that live on land are also at risk of extinction. Humans are building on areas that were once wilderness. Since 1992, urban, or city, areas around the world have more than doubled. This impacts many animals' **habitats**. It brings humans into closer contact with animals. Some animals have also adapted, or changed to live better, in urban areas.

Some species will die out if they can't adapt to their changing habitats.

The Arctic is a region that is especially at risk for habitat change. Climate change is affecting this area faster than anywhere else on the planet. Sea ice and glaciers are melting faster than expected. Polar bears need sea ice to hunt and survive. Walruses, Arctic foxes, narwhals, and seals are also affected by disappearing sea ice. If people don't take action to curb the harmful effects of global warming, many species may disappear from Earth forever.

CHAPTER SIX

FINDING FOSSIL FUELS

Humans rely on **fossil fuels** such as coal, oil, and natural gas. These fuels warm our homes, run our cars, and provide electricity. However, fossil fuels release harmful chemicals into the environment when they're burned. There's also a limited supply of these fuels. Once they've been used, they can't be replaced.

Fossil fuels can only be found deep underground. So, the only way to reach them is to mine or drill into Earth's surface. One way of drilling for fossil fuels is called **fracking**. Fracking uses a great deal of water and releases dangerous chemicals into the air and water.

People also pollute by making trash. The average American creates about 5 pounds (2.3 kg) of trash every day. Less than one quarter of this trash is recycled. The rest piles up in dumps and landfills.

There were about 2,600 active landfills in the United States in 2019.

CHAPTER SEVEN

CHANGING THE FLOW OF WATER

Every living thing on Earth needs water. However, people often change the flow of water on Earth. People build dams to block or change the flow of rivers. Some dams use the flow of water to help create electricity.

By blocking the flow of a river, dams create water shortages in some areas. This could leave plants and animals without enough water to live. Dams can also prevent fish from moving from place to place to find food and escape predators.

HOOVER DAM

The Great Pacific Garbage Patch is made up of trillions of pieces of plastic in the Pacific Ocean.

People also pollute Earth's water. More than 8 million tons (7.2 million mt) of garbage is dumped into the ocean each year. Also, dangerous chemicals make their way from factories into rivers and streams. This poisons local plants and animals.

17

CHAPTER EIGHT

CHANGING MOUNTAINS

Mountains are more than just something beautiful to look at. They're also a source of raw materials. We can find natural **resources**, such as coal, iron, and trees, in mountain environments. People have been digging mines into mountainsides for thousands of years.

Mountains are an important part of the planet's water cycle. Destroying mountain environments could create desert conditions.

Mining can change many things about a mountain, and it can be very harmful to the environment. Changing the shape of a mountain can keep water from reaching rivers and lakes. Removing trees from mountains can cause dangerous landslides. This can also destroy the natural habitats of many species. If mountains are changed or destroyed, it can affect rain or snowfall in an area. This can make an area much drier than it's naturally meant to be.

CHAPTER NINE

TIME TO ACT

Since 2017, the United States has made a number of changes to environmental laws. Lands once protected by federal law are open to activities such as logging, drilling, mining, and cattle grazing. In 2019, the federal government announced a plan to let companies drill in Alaska's Arctic National Wildlife Refuge.

Many disagree with the idea of lifting protections for the environment. As a result, many young people are stepping up to protect Earth. They feel that world leaders are failing to do so.

At age 15, **activist** Jamie Margolin and other young people decided to act. They didn't want to wait for adults to protect the environment. Margolin cofounded an organization called Zero Hour with her friend Nadia Nazar. The organization fights for climate and environmental justice.

Margolin started Zero Hour in 2017. She's shown here in 2019.

CHAPTER TEN

FIXING THE PROBLEM

Scientists and activists are working on ways to return ecosystems to their original natural state. This could mean returning native plants to an area that's been damaged or restoring natural waterways.

People can also help clean up contaminated, or polluted, areas such as the sites of oil spills. Scientists are working on new ways to do so. In 2010, bacteria were used to help clean a large oil spill in the Gulf of Mexico.

Solar panels are a way to use clean, renewable energy.

Another way to help the environment is to replace harmful actions with helpful actions. That means taking action to fix, limit, or balance out the harmful impacts that humans have on the environment. This may include recycling items or planting trees. It might mean using renewable energy sources or making existing buildings more energy **efficient**.

CHAPTER ELEVEN

GO GREEN!

Earth Day marked its 50th anniversary in 2020. The first Earth Day was organized by Wisconsin Senator Gaylord Nelson. It took place on April 22, 1970.

That day, about 20 million Americans celebrated by taking part in events calling for a healthy, **sustainable** environment. By the end of the 1970s, the United States government passed several laws that protected the environment. By Earth Day's 20th anniversary, 200 million people celebrated Earth Day in 141 countries.

GAYLORD NELSON

When he was a teenager, Jerome Foster II founded OneMillionOfUs. The organization raises awareness about climate change.

Because of Earth Day, many people became more aware of the impact they have on the planet. People started making an effort to live in a way that's less harmful to the environment. They began cutting back on the resources they use. Some started reusing and recycling items to add less trash to landfills. This environmental movement continues today.

CHAPTER TWELVE
ENVIRONMENTAL LAWS

The environmental movement of the 1970s resulted in new laws to help the planet. The Clean Air Act of 1970 helped fight the growing problem of air pollution. It gave the Environmental Protection Agency (EPA) the ability to fight environmental pollution. It set up guidelines for making air safer to breathe.

The Clean Water Act of 1972 set rules for keeping water clean and healthy. It also addressed environmental issues such as protecting water habitats.

The Endangered Species Act (ESA) of 1973 allowed the federal government to protect endangered and threatened species and their habitats. The law creates lists of plant and animal species that need protection.

However, recent changes to these environmental laws have weakened them. Some people feel that mining, drilling, and industry are now favored over the health of the planet.

William Jefferson Clinton Federal Building

12th Street, NW Entrance

U.S. Environmental Protection Agency (North and South Entrances)

The EPA deals with environmental issues in the United States.

CHAPTER THIRTEEN

GLOBAL HELP FOR THE PLANET

Environmental movements are taking place all around the world. The United Nations Environment Programme (UNEP) provides leadership about environmental issues in many places. UNEP also helps developing countries put healthy environmental practices in place.

In 2016, many countries signed the UN's Paris Agreement, the world's first global climate agreement. Its main goal is to address the human effect on climate change. Each country commited to creating and carrying out plans to reduce and fight the effects of greenhouse gases in the atmosphere. The countries also agreed to reduce their use of fossil fuels. They agreed to research new **technologies** to help reduce the amount of greenhouse gases they produce.

The United States was part of the Paris Agreement, but in 2017, U.S. President Donald Trump announced a plan to withdraw from it.

Activists around the world, many of them teenagers and young adults, continue to come together to raise awareness about environmental issues.

CHAPTER FOURTEEN

LENDING A HAND

Protecting Earth's land may feel like a lot to handle, but small changes can lead to big results. Young activists are leading the way in today's environmental movement. At age 12, Haven Coleman became a cofounder and codirector of the U.S. Youth Climate Strike. The group organizes weekly school protests. It does this to tell politicians to take action to save the environment. Activist Eyal Weintraub of Argentina was a teenager when he organized a climate protest in front of Argentina's national congress.

You can help too. You could post videos online of you and your friends cleaning a local park. You could ask your teacher if you can hold a school event to raise awareness about environmental causes. The future belongs to you. Make it a future in which Earth is protected.

GLOSSARY

activist (AK-tih-vist) Someone who acts strongly in support of or against an issue.

ecosystem (EE-koh-sih-stuhm) A natural community of living and nonliving things.

efficient (ih-FIH-shuhnt) Done in the quickest, best way possible.

environment (ihn-VIY-ruhn-muht) The natural world around us.

fossil fuel (FAH-suhl FYOOL) A fuel—such as coal, oil, or natural gas—that is formed in the earth from dead plants or animals.

fracking (FRAA-king) A drilling technique used to extract oil and natural gas from underground.

greenhouse effect (GREEN-howz ih-FEKT) The warming of Earth's atmosphere due to gases that trap energy from the sun.

greenhouse gas (GREEN-howz GASS) A gas in the atmosphere that traps energy from the sun.

habitat (HAA-buh-tat) The natural home for plants, animals, and other living things.

marine (muh-REEN) Having to do with the sea.

nutrient (NOO-tree-uhnt) Something taken in by a plant or animal that helps it grow and stay healthy.

resource (REE-sohrs) Something that can be used.

sustainable (suh-STAY-nuh-buhl) Able to last a long time.

technology (tek-NAH-luh-jee) A method that uses science to solve problems and the tools used to solve those problems.

INDEX

PRIMARY SOURCE LIST

Page 5
Plaque placed at the former location of Icelandic Okjokull glacier. Photograph. July 18, 2019. Iceland. Held by Wikimedia Commons.

Page 7
Forest fires in Brazil. Photograph. August 27, 2019. By Joao Laet. Held by Getty Images.

Page 9
Machinery buried by dust storms. Photograph. 1936. Dallas, South Dakota. Now kept at Everett Historical.

WEBSITES